# Biking for a Better Planet

## Sustainable Urban Cycling Tips and Practices

# Table of Contents

# Chapter 1. Introduction

Unleash your inner cyclist and empower yourself with every pedal you push with "Biking for a Better Planet: Sustainable Urban Cycling Tips and Practices". This Special Report is your blueprint to turning your biking passion into an eco-friendly lifestyle that is beneficial not just for your health, but for the health of our dear Mother Earth too. From in-depth guidance on choosing the right bike for city riding to savvy tips on promoting sustainability through cycling, this exciting read is sure to equip you in becoming an urban cycling ambassador. Embrace an adventure that reduces your carbon footprint while adding joy-filled mileage to your life. Doesn't that sound simply fantastic? Buy this Special Report now and start your journey soaring down the road to a greener, healthier planet. Let's ride together towards a sustainable future today!

# Chapter 2. The Joy of Cycling: More than Just an Exercise

Cycling is an activity that extends well beyond mere physical exercise. It is a form of mindfulness, a mode of transportation, a tool for exploration, and a venue for bonding with friends and family. As we delve deeper into the intricacies of urban cycling, we recognize that its joy is multi-faceted, touching upon various aspects of our lives, including physical health, mental well-being, social interactions, and connectivity with nature.

## 2.1. Physical Health and Fitness

Cycling daily could significantly enhance your overall well-being. It can cultivate cardiovascular health, increase muscle strength, boost endurance, and improve joint mobility. Studies have shown that incorporating cycling into your daily routine can lower the risk of developing chronic conditions such as heart disease, cancer, and diabetes. Cycling also bolsters fitness by pushing your heart rate up, all the while giving you control over intensity. Whether you prefer a leisurely ride or an uphill sprint, you tailors your workout to your specific needs, making it suitable for all fitness levels.

## 2.2. Mental Well-being

Pedaling across urban landscapes has tremendous benefits for mental health, too. Cycling can help alleviate feelings of stress, depression, or anxiety. The combination of exercises stimulates the release of happy hormones, such as endorphins, while the serene and rhythmic nature of cycling can have meditative effects on the mind. Taking a break from the fast-paced city life to smoothen out your pedaling rhythms can help clear mental clutter and cultivate an inner sense of peace.

## 2.3. Social Benefits

Cycling, particularly in urban settings, yields social benefits that include stronger community ties and collective consciousness towards the environment. Biking is a highly social activity – communing for rides, participating in cycling events, or joining a local cycling club can establish camaraderie with fellow biking enthusiasts. This shared interest brings people together, promoting richer social interactions, empathy, and a sense of personal identity.

# 2.4. Connectivity with Nature

Naturally, one of cycling's primary highlights is connectivity with the outdoors. As urban dwellers, we rarely get the chance to appreciate the natural beauty incorporated into our cities – the vibrant blooms in the park, the rustling leaves through the breeze, and the shimmering rivers reflecting the city skyline. Biking can reawaken the human-nature connection that is usually diluted by urban living.

# 2.5. Combining Transport and Exercise

In cities, commuting using private vehicles or public transit can become routine and tedious. Cycling presents a unique solution to incorporate exercise into your daily commute. It can transform journeys into joyful experiences filled with exhilaration, exploration, and autonomy rather than just a means to get from point A to B. By adopting cycling as a mode of transport, you'll never have to endure rush-hour traffic or crowded public transport again, instead enjoying the open air and freedom of the cycle lanes.

# 2.6. Exploring Urban Landscapes

Cycling is not merely a mode of transport but a way to explore your urban landscapes more intimately. The experience of weaving through streets and exploring undiscovered routes adds an adventurous edge to your daily routine. No car, bus, or train journey can mimic the sense of achievement and exploration that comes from discovering a new path on your bicycle. It's a true expression of our inherent curiosity, and adventurism set against the backdrop of your cityscapes.

# 2.7. The Role in Sustainability and the Environment

Perhaps the most critical aspect of cycling in the concrete jungle is its close link with sustainability. Cycling, a zero-carbon mode of transport, substantially reduces the carbon footprint, contributing to cleaner air and quieter roads. Each pedal pushed towards sustainability spearheads a wave of change that is more urgent in our urban environments than anywhere else. It isn't just an activity, but a collective step towards a greener future.

Cycling in the city holds a string of pearls with each pearl representing a joy derived from this simple activity. Engaging in this practice can open doors to a better lifestyle and a more sustainable world. The joy of cycling, therefore, is more than exercise – it is the joy of life itself.

# Chapter 3. Choosing Your Ride: Comprehensive Bike Buying Guide

Congratulations on taking the first step toward a more sustainable future with urban cycling. Picking the right bicycle for city commuting might roughly feel akin to navigating a maze, especially if you're a beginner. Don't worry, with this comprehensive guide, we will simplify the process and help you make an informed decision. In this chapter, you'll learn about the broad spectrum of bikes available in the market, consider some key aspects when buying a bike and explore essential accessories that'll make your journey smoother and safer.

## 3.1. Know Your Bikes

Before you can choose the right bike, understanding the different types (also known as "geometry") of bikes is essential. There are several types designed for specific kinds of terrain, riding styles, and preferences.

### 3.1.1. Road Bikes

Designed for speed, road bikes are perfect for long-distance highway rides and racing. They have thin, slick tires for cutting the wind, and the light, aerodynamic frames make acceleration and climbing hills easier. However, they might not be the best choice for city commuting due to their aggressive riding position making it less comfortable for everyday use, and thin tires not best suited for navigating through rough urban terrains like potholes.

## 3.1.2. Hybrid Bikes

A mix of road and mountain bikes, hybrid bikes are designed with comfort and functionality in mind, making them suitable for city commuting. They typically have thicker tires compared to road bikes, meaning they can better take on debris-strewn city streets. Their upright position is easier on the back and makes for better visibility in traffic.

## 3.1.3. Mountain Bikes

Mountain bikes are designed for rough, off-road trails. They have rugged frames and wide, durable tires to cushion against rocks, roots, and bumps as you navigate across uneven paths. While they provide excellent comfort, their bulky design and slower speed might not be ideal for an everyday city commute.

## 3.1.4. Touring Bikes

Touring bikes are designed for carrying loads and for long-distance travel. They have sturdy frames capable of managing heavy weight and comfortably designed for long hours of pedaling. Keep in mind, though, these bikes might be overkill if you're not planning on bike tours or carrying heavy loads daily.

## 3.1.5. Cruiser Bikes

Cruiser bikes, or "city bikes," are designed for short, leisurely rides around the city. With wider tires, an upright riding position, and often a larger saddle, they're an ideal choice for a comfortable, relaxed commute. However, they're not intended for hilly terrains or long trips due to their heavy weight and inability to handle significant speed or distance.

# 3.2. Gauge Your Preferences

Once you have a basic understanding of the bike types, the next step is to identify your needs and preferences. Here are the factors you need to consider:

## 3.2.1. Riding Surface

Different bike designs handle different environments better. Road bikes are best on smooth pavement, mountain bikes are meant for rough, off-road terrains, while hybrid bikes can handle a decent mix of both. Carefully consider where most of your riding will be.

## 3.2.2. Distance

If you're planning to ride longer distances regularly, a lightweight, aerodynamic bike like a road bike or a touring bike could make the journey much easier. For short trips, you might find cruiser or a hybrid bikes more comfortable and convenient.

## 3.2.3. Seating Position

The curvature of a bike's handlebar affects your seating position and thus, comfort. Mountain, hybrid, and cruiser bikes allow for a more upright ride. Road and touring bikes, on the other hand, require a hunched over position to maximize speed but could be less comfortable.

# 3.3. Try Before You Buy

Once you've narrowed down your options, visit a local bike shop to take some test rides. Consider the following before and during your test ride:

### 3.3.1. Bike Sizing

Though bikes come in standard sizes (small, medium, large), what fits you best may depend on the particular brand or model. A bike that's too large will be hard to control, while a bike that's too small will feel cramped. Be sure to consult with the shop's personnel to get the best fit.

### 3.3.2. Suspension

Mountain and some hybrid bikes offer suspension to reduce impact on rough terrains. The downside is that it can make pedaling less efficient by absorbing some energy. Consider this if you ride on bumpy roads often.

### 3.3.3. Gear Range

Also called the "speed" of the bike. It refers to the number of gears the bike has. More gears provide flexibility for different terrains but might complicate your ride if your city commute is relatively flat.

# 3.4. Essential Bike Accessories

Finally, remember that a bicycle isn't complete without its accessories. They add comfort, safety, and convenience to your ride.

### 3.4.1. Lights

A front light and rear light are essential for riding in dim or dark conditions. They enhance your visibility to motorists.

### 3.4.2. Helmets

Never compromise safety. Helmets are a must-have for every cyclist.

### 3.4.3. Locks

If you're going to be parking your bicycle in public places, invest in a good lock for theft prevention.

### 3.4.4. Mudguards and Racks

Install mudguards to protect you from road splash. A rack is useful if you plan on carrying loads regularly.

With this comprehensive guide, you're all set to choose the best bike for your city cycling needs. Remember, comfort and functionality should be your focal points when selecting your ride. Happy cycling, and here's to a greener, healthier planet!

# Chapter 4. Safety First: Essential Gear and Equipment for Urban Cyclists

An important aspect of riding your bicycle in an urban environment is understanding the appropriate safety gear and equipment that you need. Your personal safety is paramount – and equipping yourself correctly is one part of your first line of defense against potential mishaps. Getting the right equipment will not only protect you but also optimize your cycling experience.

## 4.1. Choosing the Right Helmet

One of the first pieces of safety gear every cyclist needs is a helmet. Choose one that is certified, well-ventilated, fits well, and comfortable to wear. Keep in mind that if a helmet does not fit properly, it will not function as intended. Look for a sticker that confirms the helmet meets safety standards and tests. As an urban cyclist, an important feature to consider is the helmet's visibility in traffic. Choose bright colors or helmets with reflective patches that make you easily visible to drivers.

## 4.2. Importance of Appropriate Cycling Clothing

Efficient cycling isn't just about the bike, your clothing can make a big difference. General rules to follow when choosing cycling clothing include avoiding loose clothes that could get tangled in your bike, and wearing bright or reflective colors for visibility. Always opt for moisture-wicking and breathable material.

## 4.3. Shoes with the Right Grip

Wearing the correct shoes is vital when cycling in an urban environment. Look for a pair that offers good support, has a robust sole, and is designed to grip the pedal. Some cyclists choose specially designed cycling shoes that clip onto the pedals, offering greater control.

# 4.4. Gloves for Comfort and Protection

Cycling gloves protect your hands in falls and absorb vibrations that could lead to discomfort or fatigue. They can also improve your grip on the handlebars. Be sure to choose gloves that fit well and don't restrict movement.

# 4.5. Eye Protection

Glasses protect your eyes from dirt, dust, and flying debris as well as from wind and the sun's UV rays. Choose lenses that provide UV protection and frames that are comfortable, preferably with vents to prevent fogging.

# 4.6. Reflective Gear and Lights

One of the biggest risks for cyclists, particularly in urban environments, is not being seen. Therefore, it's important to invest in reflective gear and lights. A high-visibility vest or jacket can greatly enhance your visibility. Additionally, mounting lights on your bicycle is a necessity, not a luxury. They also serve a dual purpose: helping you see the road in dark conditions and enabling other road users to see you.

## 4.7. Comfortable and Robust Saddle

A comfortable saddle is essential for enjoyment and endurance. Look for a saddle that fits your body type and is supportive. A poorly chosen saddle can lead to discomfort and chafing over distances.

## 4.8. Addition of Mirrors

Mirrors aren't just for motor vehicles. Bicycle mirrors give you a view of the traffic behind, essential in busy urban environments. Some bike helmets even come with mirrors attached.

## 4.9. Regular Bicycle Maintenance

A bicycle in good working order is also an important piece of safety equipment. Regular maintenance can prevent many problems from occurring during your commute. Check your brakes and cables regularly, keep your chain clean and lubricated, and ensure your gears are shifting properly.

## 4.10. Carrying Water and a Repair Kit

Don't forget to always carry a water bottle to stay hydrated and a basic repair kit. The latter should contain a spare tube, a set of tire levers, a multi-tool with basic wrenches and screwdrivers, and a patch kit.

## 4.11. Portable Locks for Security

Finally, an often overlooked, but essential piece of cycling equipment in an urban environment is a good quality bicycle lock. Make sure it's robust and portable. The last thing you want is your bicycle being stolen when you've parked it.

Understanding and investing in the right gear and equipment can make your urban cycling experience safer, more comfortable, and enjoyable. So gear up, practice your riding skills, get familiar with your route, and keep these safety considerations in mind to enjoy a greener ride around the city.

# Chapter 5. Mastering City Terrain: Tactics for Safe and Efficient Biking

Urban environments provide unique challenges for cyclists. Skyscrapers, busy streets, crowded sidewalks, and constantly changing traffic patterns can all make navigating a city on bike a daunting task. However, with the right tactics and techniques, you can master city terrain and bike safely and efficiently.

## 5.1. Understanding Urban Environment

The first step in mastering city terrain is understanding the unique challenges it presents. Cities are characterized by congested traffic, unexpected obstacles, and varied road conditions. As a cyclist, you will need to navigate through these challenges smoothly.

## 5.2. Mapping Your Route

Before hitting the streets, it's essential to map out your route. Use a map to find the most bike-friendly routes in your city. Some cities have dedicated bike lanes, while others have shared roads. Look for routes that are flat, wide, and have well-maintained roads. Remember that the shortest route is not necessarily the fastest or safest.

## 5.3. Observe and Adapt

Alertness and adaptability are key in mastering city terrain. Stay aware of your surroundings, including moving vehicles, pedestrians, and other cyclists. Be prepared for sudden changes in traffic patterns and road conditions. Watch out for obstacles like potholes, loose gravel, and debris on the road.

## 5.4. Cycling in Traffic

Negotiating traffic can be tricky, but there are strategies to keep you safe. Always ride in the same direction as traffic. Claim your lane instead of hugging the curb; it makes you more visible to drivers and offers space to maneuver. Remember that traffic laws apply to cyclists, too.

## 5.5. Negotiating Intersections

Intersections can be dangerous for cyclists. Approach with caution and make eye contact with drivers to ensure they see you. When turning left, be sure to signal your intention and move to the center of your lane well before the intersection.

## 5.6. Bike Maintenance

A well-maintained bike is a safe bike. Regularly check brakes, tires, and lights to ensure they are in good order. Consider investing in a bike repair kit to fix minor issues on the go.

## 5.7. Equipment and Gear

Proper gear enhances safety and comfort. Helmets are a must for safety, and cycling gloves can offer increased grip and comfort. Lights and reflective clothing increase visibility, particularly in low light conditions.

With these tactics and practices, you can navigate city terrain safely and efficiently. By understanding the challenges of the urban environment, planning appropriately, maintaining alertness, observing traffic laws, and caring for your bike and gear, you can transform your city cycling experience. It's not just about getting from point A to B but also about enjoying the journey and doing your bit for a greener, healthier planet.

Remember, the journey on the road to cycling proficiency might be challenging, but it's also rewarding. With every pedal push, you're not merely traversing city terrain; you're paving the path towards a sustainable urban lifestyle. As you improve, the city's frenzied streets and bustling highways will cease to be daunting and become your very own adventurous playground. Embrace the ride!

# Chapter 6. Sharing the Road: Etiquettes and Laws Every Cyclist Should Know

Cycling responsibly is not only about understanding and obeying street laws, but also about mindfulness in sharing the road with others. As cyclists, it is our duty to maintain safety, adhere to cycling essentials, and foster a supportive environment for everyone on the road.

## 6.1. Understanding the Road Rules

To effectively share the road, every cyclist must have a thorough understanding of the road rules. These rules may vary by location, so it's crucial to familiarize yourself with local regulations, but here are some universal principles:

- Always ride in the same direction as the traffic. Remember, cyclists are considered part of the traffic and should abide by the same rules.

- Use cycling paths when they are available.

- When cycling paths are not available, stick to the far right side of the roadway whenever possible.

- Follow all traffic signals and signs, such as red lights and stop signs. Treat these signals the same way as if you were driving a car.

# 6.2. Cyclist Signals: Ensuring Your Intentions Are Clear

As part of obeying traffic rules, cyclists should also learn a set of hand signals. This silent communication is critical in maintaining safety on the road and informing both drivers and pedestrians about your intentions. Here are what each hand signal generally means:

- Left arm outstretched horizontally: Indicates you plan to turn left.

- Right arm outstretched horizontally or left arm at 90-degree angle upwards: signifies a right turn.

- Left or right arm pointing downwards: Shows you plan to stop or decrease your speed.

Always make sure to perform these signals in a manner that's clear and visible to others around you.

# 6.3. Rights and Responsibilities of a Cyclist

As cyclists, we have certain rights and responsibilities on the road. Understanding these is crucial for forging a harmonious relationship with other road users. Some of the key points include:

- Right to the road: Cyclists have the same road rights as any other vehicle.

- Stay to the right: Unless you are turning left, passing another vehicle, or avoiding hazards, it's advisable to stay on the right of the roadway.

- Yield to pedestrians: In many regions, it's the law to yield to pedestrians on sidewalks and crosswalks.

- Respect traffic signals: Just like motorists, cyclists must adhere to all traffic signals and signs.

## 6.4. Promoting and Practicing Road Etiquette

Promoting and practicing road etiquette are honorable actions that set the tone for respectful and safe behaviors. They serve as moral guidelines for cyclists and help improve how cyclists are perceived by other road users.

- Stay observant: Be aware of your surroundings at all times. This includes being mindful of the traffic, pedestrians, and any potential obstacles on the road.

- Respect and kindness: Treat everyone you encounter on the road with respect and kindness. One small act of courtesy can inspire others to do the same.

- Maintain your bike: Keep your bike in good condition. Broken lights, faulty brakes, or a misaligned handling bar can lead to accidents.

- Do not obstruct traffic: While we have a right to the road, we shouldn't misuse it and cause traffic. Always remember that roads are for everyone to share.

## 6.5. Encouraging Responsible Road Behaviors

Being a responsible cyclist can translate to influencing others—both cyclists and drivers—to adopt responsible road behaviors.

- Safe road practices: Encourage your fellow cyclists to learn and follow safety precautions and road laws.

- Wearing helmets: Savvy cyclists wear helmets regardless of the legal requirements. A helmet protects your head in case of accidents. You also encourage others to do the same by wearing one.

- No distractions: Using mobile phones or listening to loud music while cycling can prove fatal as these distractions can lead to accidents. Inspire others by sharing these facts.

- Sustainable cycling: Advocate for cycling as a sustainable travel option. It encourages both healthy and eco-friendly habits.

In conclusion, sharing the road requires both legal compliance and moral responsibility from cyclists. By understanding the road laws, performing appropriate cycling signals, practicing road etiquette, and promoting responsible behaviors, we can pave the way for safer and more harmonious road sharing experiences. This, combined with our daily biking, makes us true ambassadors for urban cycling, paving the way for a more sustainable future.

# Chapter 7. Maintenance Matters: Keeping Your Bike in Top Shape

The maintenance of your bicycle plays a critical role not just in ensuring smooth and trouble-free rides, but also in extending the lifespan of your bike and reducing waste. Tailoring a routine that includes regular check-ups and proactive practices can be a game-changer in maintaining optimal performance of your two-wheeler. But worry not, we are here to guide you through this process.

## 7.1. Become Your Own Mechanic: Basic Bike Anatomy

Understanding the basic components of your bike provides a great foundation for maintenance. So, let's unpack some crucial bike parts you should be familiar with:

1. Chain: Transfers power from the pedals to the wheel

2. Gears: Allow you to adjust the level of resistance when pedaling

3. Brakes: Control the speed and bring the bike to halt

4. Tires: Facilitate contact with the ground

5. Pedals: Where the rider places foot for propulsion

6. Handlebar: Provides the rider with steering control

Armed with this knowledge, you are now better equipped to deal with routine maintenance and potential issues that might arise.

## 7.2. Keeping it Clean: The Importance of a Spotless Bike

Cycling in urban environments exposes your bike to a myriad of factors that can lead to grim and grit accumulation. While it may seem harmless, this can actually have adverse effects on the functionality of your bike over time. A clean bike not only functions optimally but also allows for easier detection of potential issues in their early stages.

You should strive to keep your bike as clean as possible. Use a biodegradable cleaning solution and a brush for effective results. Focus on the chain and gears, as they may gather more dirt compared to other parts.

## 7.3. Ride Smoothly: Lubrication Fundamentals

Lubrication is a vital part of bike care. It reduces mechanical friction, protects against rust, and consequently, prolongs your bike's lifespan. Pay particular attention to the chain, derailleurs, brake and shifter pivots, ensuring they are well-lubricated.

However, avoid overlubrication as excess lubricant tends to attract dust and debris. After applying lubricant, allow it to penetrate into the necessary parts then wipe away any excess.

## 7.4. Check Your Treads: Tire Care and Maintenance

Tire integrity is pivotal for safe and efficient riding. Regularly check your tires for punctures, embedded object, wear, and pressure.

The tire pressure greatly affects your ride. A poorly inflated tire can increase rolling resistance and make your ride more strenuous. Moreover, it increases the risk of punctures. Know the recommended tire pressure for your bike and periodically, using a proper gauge, ensure it's within the suggested range.

## 7.5. Braking it Down: Brake Performances and Upkeep

Monitoring your brake system is essential to ensure safety during your rides. Make sure your brake pads don't show excessive wear. If the brake levers touch the handlebars before maximum resistance is felt, then your brake cables could be too loose and might require tightening.

Remember that your safety and that of other road users come first so always ensure your bike's brake system is faultless before you venture out on the road.

## 7.6. Listen to Your Bike: Spotting and Addressing Noises

Squeaking, creaking, or grinding noises are your bike's way of crying out for your attention. These could indicate a variety of potential issues - from a loose component, lack of lubrication, to even misalignments. Address these issues immediately to prevent them from escalating.

## 7.7. Regular Check-ups: The Value of Consistent Maintenance Routines

While it's important to react to issues as they arise, proactive maintenance is the key to a long-lasting bike. Regular check-ups will help identify any underlying issues that might otherwise go unnoticed until it's too late to resolve them simply.

Bike maintenance is not a complex task but it does demand time and commitment. However, the benefits far outweigh the effort. Regular bike care ensures smoother and safer rides, extends your bike's life, and, in the long run, contributes to sustainability by reducing waste and the need for replacement parts. Take care of your bike, and it will take care of you on your journey to a greener and healthier planet. Pay attention to each pedal you push, because each rotation can be a stride towards sustainability if your bike is properly maintained.

Remember, every revolution begins with a single push. A well-maintained bike lends an empowered push towards greener cities. Start today.

# Chapter 8. The Green Effect: Understanding the Environmental Impact of Cycling

In an era of increasing environmental challenges and crises, including climate change, pollution, and the brunt of non-renewable resource exploitation, exploring greener lifestyle alternatives is not just beneficial, but a necessary choice. Cycling, as an activity, possesses tremendous potential for sustaining and improving the health of our environment, primarily due to its lower carbon footprint. A meticulous understanding of this impact of biking serves as the first step in our endeavor to establish a greener lifestyle.

## 8.1. Embracing Carbon-free Transportation

In today's urbanized society, transportation is the second-largest source of carbon dioxide ($CO_2$) emissions, constituting about 29% of the total. On the contrary, cycling is inherently climate-neutral. Being human-powered, it does not consume fuel or emit any harmful emissions directly. Each kilometer traversed by bicycle instead of a car prevents approximately 0.15 kilogram of $CO_2$ emissions. Over a year, an average commuter changing the mode of transport from car to bicycle could potentially save around one ton of $CO_2$.

## 8.2. Resource Efficiency and Cycling

Cycling is not just an eco-friendly transport method due to zero direct emissions; it's also incredibly resource-efficient. Bicycles weigh about

20 times less than the average car, resulting in less energy required to propel them forward. While car production consumes considerable resources (materials, energy), bicycles are much simpler machines, with fewer components requiring smaller amounts of raw materials. Additionally, with zero reliance on fuel or electric charging, cycling outperforms most transport modes in operational energy consumption.

# 8.3. Preventing Air and Noise Pollution

Air pollution is a significant concern in various urban areas worldwide. Bicycles produce no exhaust fumes, contributing to cleaner, healthier air. Furthermore, bikes generate low levels of noise, fostering quieter, relatively tranquil urban environments.

# 8.4. Enhancing Urban Green Spaces

Creating infrastructures like cycling lanes and bike parks often blends naturally with the development of green spaces in cities, which has numerous environmental benefits, including enhancing biodiversity, improving air quality, and offering recreational value. Additionally, these green spaces provide intrinsic ecosystem services, such as stormwater management and temperature regulation.

# 8.5. Impact on Public Health and Wellbeing

Health impacts aren't often considered part of environmental footprint analysis, but they should be. Cycling leads to healthier lifestyles that drastically reduce the incidence of chronic illnesses like heart disease and diabetes, which in turn, diminish the dependence on healthcare services. Fewer demands on the healthcare system mean less resource extraction, energy consumption, waste production, and CO2 emissions, linking back to better environmental health.

## 8.6. The Lifecycle Analysis

A "cradle-to-grave" analysis of bicycles highlights its much lower environmental implications when compared to motor vehicles. This analysis includes the resource requirements and emissions created during the manufacturing, use, and disposal phases of the bike's life. Bicycles outperform motor vehicles on nearly all these scales, reinforcing their green credentials.

## 8.7. Encouraging Sustainable Economic Development

Apart from environmental benefits, cycling promotes a sustainable economic model. Cities investing in extensive cycling infrastructure see improvement in local tourism, reduced health care costs due to healthier citizens, unemployment reduction through bike-related job creation, and overall boosts in the economy.

## 8.8. Cycling and Environmentally Responsible Behavior

Incorporating cycling into one's lifestyle may foster environmental awareness and overall behavior change. By biking, people become more connected to their local environment, fostering a sense of responsibility and sparking motivation to make additional changes in their lives that further reduce their environmental impact.

Understanding these areas of environmental impact is vital to fully grasping the power of cycling for healthier, sustainable urban living.

While it's not feasible for everyone to trade a car for a bicycle completely, integrating cycling more into our lives, even if just for shorter trips or recreational purposes, can lead to considerable environmental gains. Let's pedal towards a future where nature and urban life are balanced, and our environmental responsibility and consciousness are heightened.

# Chapter 9. Biking with a Purpose: Incorporating Cycling into Your Daily Routine

We often get caught up in the daily grind of life, commuting from point A to point B using gas-guzzling and pollution-promoting automobiles. But imagine, if you will, incorporating a lifestyle shift that achieves your necessary daily commute while simultaneously benefiting your health, reducing transportation costs, and most significantly - positively impacting the environment. Cycling, a relatively simple and accessible activity, presents this opportunity. The extravagant charm of this beautiful two-wheeler isn't just its affordability or proven health advantages, but also the way it contributes to a sustainable lifestyle, enabling us to reduce our carbon footprint, embrace an eco-conscious lifestyle, and 'bike with a purpose'.

## 9.1. The Morning Commute: Biking to Work

Embarking on your everyday journey using a bicycle can turn your mundane route into an invigorating start to the day. Cycling to work has the power to transform your entire experience of a working day. Apart from the much-touted health benefits that include improved cardiovascular and lung health, it helps in reducing stress levels and promoting mental wellbeing.

Before you begin, do proper research about the safest and most bike-friendly routes in your city. Ensure that it is safe and secure to park your bike at your workplace. You might need to wake up a bit earlier to accommodate the added commute time, but soon enough, the satisfaction of reducing pollution, saving money, and getting exercise would make it worth the change.

## 9.2. Running Errands just Got Healthier

Essential tasks like going to the grocery store or dropping off mail can become a part of your daily exercise when incorporated with cycling. One of the best parts of using a bicycle for such errands is leveraging 'dead time'. Resist the urge to hop into your car for quick rides to your neighborhood store, instead, put your cycling gear on. Install baskets or panniers on your bike to hold your purchases. Over time, these short rides will add up to significant mileage and energy expenditure.

## 9.3. The School Run

Convincing your child to cycle can have far-reaching benefits - it lays the groundwork for a healthier lifestyle from a young age, encourages independence, and inculcates the principles of sustainability. For younger children, a family ride to school might be a feasible option, while older kids can be encouraged to cycle independently or in groups with neighborhood friends. This act promotes a sense of responsibility and time management among the young riders.

## 9.4. The Delightful Blend – Combining Public Transport with Cycling

Sometimes, distances are too great or time too short to be covered just by cycling. In such instances, combining cycling with public transportation is an excellent compromise. Bicycles can be taken aboard buses and trains, allowing you to cycle the initial and final parts of the journey and cover the longer, more time-consuming parts using public transport. This combined mode of travel allows you to maintain an active lifestyle while reducing carbon emissions.

## 9.5. Leisure Time Cycling

Leisure rides can provide a welcome break from a monotonous daily routine, provide a chance to connect with nature, and offer an opportunity to explore the local scenery and attractions. Planning ahead and identifying cycle-friendly leisure spots boost the prospect of making cycling a fulfilling pastime and not just a means to commute.

## 9.6. Preparations and Planning

When deciding to integrate cycling into your daily routine, it's essential to plan well. Initial steps involve procuring a suitable bike, necessary safety gear, and a suitable and secure method of storing the bike at each destination. Familiarize yourself with local traffic rules and regulations and identify the most bike-friendly paths to each of your regular destinations.

In summary, 'biking with a purpose' is an endeavor that goes beyond personal health and extends to the wellbeing of our planet. Adopting an integrated cycling routine is about being a part of the solution to environmental issues and participating in the creation of sustainable cities. The shift may take some adjustment, but the benefits will foster a harmonious relationship with our surroundings – a stride that will make all the difference. It's time we pedal our way into a healthier and sustainable future, one bike ride at a time.

# Chapter 10. Strengthening the Cycling Community: Building Bridges with Local Groups and Events

As an enthusiastic advocate for bicycling, you have the potential to make a profound impact in your community. With your passion and knowledge, you can become an ambassador for the benefits of cycling—whether it's for personal fitness, as an alternative mode of transportation, or for environmental reasons. It all comes down to initiating conversations, encouraging participation, and supporting the local cycling community.

## 10.1. Engage with the Cycling Community

Engaging with the cycling community is one of the most effective ways to promote the lifestyle and ethos of biking. Attending bike rides, meetings, seminars, and workshops can provide a fantastic platform to share your experiences and learn from others. Be sure to participate regularly to foster lasting relationships.

Participate in Local Bike Rides: Join community bike rides to make friends with kindred spirits who share a passion for cycling. This is also a perfect avenue to inspire wellness, environmental awareness, and bicycle safety.

Attend Meetings and Seminars: Get to know the movers and shakers within local urban planning and traffic safety committees, where you can voice your thoughts and concerns related to cycling. This is a valuable method to contribute to the community by advocating for bike-friendly changes in local policies and planning.

Join Workshops: Don't overlook the numerous workshops offered by some cycling groups. These platform ranges from bike maintenance to safe cycling practices, and even novice classes for beginners. Improve your skills and share your knowledge along the way.

## 10.2. Foster the Cycling Culture

Promoting a cycling culture goes beyond encouraging your friends and family to take up biking. As a committed cyclist, you can set an example and inspire a larger audience.

Start a Local Cycling Club: If there's no local cycling community, consider starting one. This can be a fun and effective way to gather cycling enthusiasts. You can even forge partnerships with local businesses to sponsor events or rides.

Promote Cycling to Work/School: If feasible, encourage your coworkers, friends, and family to bike to work or school. You can also engage local schools and organizations to regenerate interest in biking and educate the community about its benefits.

Encourage Bike-Friendly Businesses: Propose bike racks in popular locations like cafes, parks, shopping centers, and restaurants. Advocate for businesses to offer discounts or incentives to customers who arrive on bikes.

## 10.3. Collaborating with Local Government and Non-Profit Organizations

An essential element to strengthening the cycling community is to collaborate with local government bodies and non-profit organizations. Push for bike-friendly infrastructure and safety protocols.

Advocate for a Better Infrastructure: Canvass for bike lanes, bike racks, bike-friendly traffic signals, and other infrastructure enhancements. They not only encourage bicycling but also make it safer and easier.

Promote Safe Cycling Practices: Use your platform to spread awareness about traffic laws, safety gear, and safe cycling habits. Organize workshops or seminars to educate both cyclists and non-cyclists alike.

Collaborate with Non-Profit Organizations: Establishing partnerships with non-profit organizations that promote bicycling can magnify your impact. Funding, resources, and a more extensive network can do wonders for your mission.

# 10.4. Organize and Participate in Events

Organizing events centered around cycling is an excellent way to rally support for the cause.

Community Bike Rides: Whether it's a leisurely ride on a weekend, an adventurous trail ride, or a bike-a-thon for charity, group rides foster camaraderie and promote cycling.

Cycling Competitions: Engaging competitive spirits could be another route to bolster interest. Organize races or time trials to light the competitive spark within the community members.

Bike Maintenance Clinics: Frequent clinics where people can learn the basics of maintaining their bikes can advocate self-sufficiency and better care for their bikes, prolonging their bikes' lives, which in turn promotes sustainability.

In the end, strengthening your community via cycling is about more than improving health or reducing emissions—it's about fostering relationships and driving change. With every person inspired, every bridge built, and every pedal pushed, you make urban cycling—and the world—better. Don't forget, the most significant change often comes from local, grassroots efforts by folks like us. So, let's make the most out of our efforts and keep cycling towards a better tomorrow!

# Chapter 11. Advocating for the Future: Promoting Sustainable Practices in Urban Planning

As urban sprawl continues unabated, we are grappling with foreseeable issues that could hamper sustainability in our cities. Fortunately, integrating eco-friendly and health-promoting practices into urban planning can significantly reverse these concerns. By encouraging urban settlers to take up sustainable methods like cycling, we can reduce pollution, minimize traffic congestion, promote health and wellness, and rejuvenate the local economy. In this pursuit, every pedal counts. Let us delve deeper into understanding how we can promote sustainable practices in urban planning and stand up for a greener future.

## 11.1. The Eco-friendly Advantage of Cycling

When you choose to cycle, you're taking one giant leap towards environmental preservation. Unlike motor vehicles that depend heavily on fossil fuels, bicycles utilize human pedal power, thereby generating zero harmful emissions. In fact, a single car commuter switching to cycling can reduce their carbon footprint from daily commutes by at least 6 tonnes each year!

In addition, cycling can make an impact on urban air and noise pollution levels. Illustratively, a city adopting a 20% cycling share in its transport ideas can lower its noise pollution by 1 decibel and can reduce life years lost due to air pollution by approximately 30%.

## 11.2. The Role of Urban Planning in Promoting Cycling

Urban planning plays a pivotal role in promoting sustainable practices such as cycling. The layout and design of a city greatly influence the choices and lifestyle habits of its dwellers. Therefore, accommodating facilities like dedicated bike lanes, bike-sharing platforms, and cycling-friendly infrastructure can stimulate more city dwellers to adopt cycling as a key part of their daily lives.

By emphasizing on creating wide, well-marked cycling paths alongside main roads and adopting a comprehensive traffic calming policy, authorities can encourage more residents to swap their car keys for a bicycle helmet. Further, the integration of cycling routes with public transportation can support those with longer travel distances, making cycling more accessible to a broader population.

## 11.3. Community Engagement and Awareness

Engaging the community through creating awareness is key to fostering a culture that values and understands the benefits of sustainable practices like cycling. Organizing cycle tours, bike rallies, workshops, and events on eco-friendly living can build deeper ties within communities and prompt people to consider cycling as a viable mode of transport. Furthermore, teaching about the impact of cycling on the planet's health in schools and community centers can seed the idea of a greener lifestyle in the minds of the younger generation.

## 11.4. Encouraging Cycling in Business Districts

Promoting cycling in business districts can tremendously curb air pollution and traffic congestion. Businesses can play an instrumental role by providing incentives to employees who cycle to work. By installing bicycle parking spots and shower facilities, companies can facilitate their employees to cycle without worries about safety and hygiene. Furthermore, businesses can also sponsor city bike-sharing programs, strengthening their reputation as eco-friendly establishments.

## 11.5. Building Partnerships with Local Governments

Cycling enthusiasts and sustainability advocates must voice their concerns and ideas to local governments. Collaborative planning can lead to the creation of safer cycling routes, improved infrastructure, and bicycling laws that protect cyclists. By actively partnering with local governments, we can ensure better planning and proper execution of sustainable practices at a community level.

## 11.6. Fostering a Sustainable Economy through Cycling

The economic benefits of cycling cannot be understated. When individuals choose to cycle, they save on gas, maintenance, and parking, which directly impacts their wallet positively. Cities that encourage cycling attract cycle tourists and are more likely to have flourishing local businesses due to the slowed pace of travel.

Moreover, bicycle production and maintenance is a sector that provides numerous jobs, contributing to local employment. As cycling grows in popularity, so does the demand for bikes and bike-related products and services. This leads to the proliferation of local bike shops and repair businesses, stimulating the local economy.

In conclusion, promoting sustainable practices through cycling brings about a myriad of intertwined benefits. It is not just about reducing our carbon footprint and fighting climate pollution. It's about improving public health, fostering a sense of community, supporting the local economy, and augmenting the liveability of our cities. The challenge of change lies ahead of us, but as we pedal into the future, the vision of a sustainable urban environment becomes all the more achievable. Let's take the pledge to incorporate cycling into our lifestyle, advocate for supportive policies, and build our cities with sustainable mobility at the forefront. The future of our planet hangs in balance and it's time we tilt it towards sustainability.

www.ingramcontent.com/pod-product-compliance
Lightning Source LLC
Chambersburg PA
CBHW071012260726
48661CB00007B/2921